Honest Miracle + Kim Calera

Zero Waste Secrets:

The Ultimate Guide Book For A Realistic Zero Waste Lifestyle

Contents

Introduction

If you're reading this book you're probably keen on saving our beautiful planet from waste and toxins. You might have also experienced what I call "zero waste overwhelm" where you want to do everything to fight climate change, heal the planet and tell everyone about it to get them involved too! That's brilliant, but we all have our limits, so don't push yourself to do too much at once. I like to push myself just passed my limit then wait there for a little while before trying to get to the next level because if we burnout we can't help anyone.

The top excuse I've heard from people as to why they don't go greener or try zero waste is that they can't be perfect like the social media influencers who fit all of their years rubbish into one jar. We don't have to be that extreme! In fact, how do we know they actually did fit all of their years rubbish into one jar?! There's a thought...

So, I created this book as a realistic guide to zero waste, maybe that's not the best name, maybe it should be called conscious living, or less waste, because in all honesty, it's impossible to have ZERO waste.

My hope is that you will use this guide to ease your way into zero waste/less waste/conscious living, or you can use it to add to your current eco knowledge. Don't forget to share it with family and friends, so that we can all do our bit to help the planet and to also help our families and ourselves. Because we're all in this together!

Kitchen

First of all, only buy what you need. It sounds simple but this means always going through your fridge before you go food shopping, thinking about how many meals you'll be eating at home for the week and how many people you have to cook for. Then, make a list and stick to it! The average UK household throws away 22% of their weekly food shop which adds up to £700 per year!

In order to tackle climate change in your kitchen, thinking about your shopping habits is also crucial. This means taking an interest in where your produce comes from, how it got to your supermarket shelf and what packaging it comes in. Our way of shopping has become an increasingly wasteful process.

Fuelled by the modern-day expectation that we can have whatever we want, whenever we want, supermarkets stock produce all year round despite it being out of season. To do this, they ship fruit and vegetables from all over the world which produces unnecessary carbon footprint and contributes to global warming.

Plastic also features heavily in this process. Supermarkets seem to have significantly increased the plastic they wrap fruit and vegetables in, which produces huge amounts of waste. By choosing locally grown produce that comes loose (not wrapped in plastic) you can cut down on your waste significantly and start shopping zero-waste style.

Learn how to recycle properly. Look for the recycling labels on any packets before you buy and choose the packaging from materials that you know you can recycle easily. Try to support companies that use packaging made from at least partly recycled materials, but biodegradable is always best. Mixed materials often make packaging harder to recycle, so simple packaging is generally best. Be aware that often labels on the packaging are misleading. Just because it says it can't be recycled, doesn't mean to say it can't be recycled in your area, so try to learn more about what your area can and can't recycle. Local sustainability groups are useful sources of information for this and most councils have comprehensive guides on their websites for you to find out what to recycle where.

It does help to rinse out your tins, bottles and jars. Because we are always going to need more produce, we will need more tins, bottles, jars, cardboard. So the best thing for produce you can't get unpackaged or in biodegradable packaging is to make sure that every bit of packaging gets to the recycling.

For cardboard packaging that is contaminated with a lot of food, the best place for it is the compost heap, but only discard the bits of cardboard that are messy with food. You can probably recycle 85% of a pizza box if you tear off the messy bits and recycle the rest. Packaging factories want every fibre back as each of those fibres that make up your cardboard box can be reused between 10 and 20 times.

Avoid putting your food waste and those messy bits of card in the general waste bin. General waste that is contaminated with food is about 3 times more costly to deal with than general waste that doesn't contain food. If your council provides you with a food waste collection service it is still ok to choose to compost at home, but do use your food waste bins for anything you don't like to add to your compost heap (like meat and fish bones or cooked food). Even if the only food waste you have is tea bags it is still worth using your food waste collection rather than putting them in your bin.

Uses For Leftovers

Bread

Bread should never be wasted. If you look at good cooking around the world, bread is often used stale and transformed into all sorts of tasty things, such as the classic Italian tomato and bread soup, pappa al pomodoro, or panzanella (tomato and bread salad). Turn already-stale bread into croutons or, if you know you won't finish your loaf before it turns, thinly slice it and leave to go stale for perfectly crisp toast – it is a fantastic vehicle for tapenade or bruschetta toppings. And you can always blitz stale bread into bread-crumbs for crispy coatings on fishcakes or to toast and scatter over pasta dishes.

Vegetables and fruit

Why not use the leftover vegetables from your Sunday roast to make pakoras? Any vegetable you have is fine. Thinly slice, add chopped onion, ginger, garlic, whatever spices you like and chickpea flour (about one-fifth of the total), make into balls in your hand, then deep-fry until crispy.

You could dry vegetable peelings to make into flavoured salts. It's really simple: put the peelings on a rack in a really low temperature oven – a 50C degrees – leave for three hours, blend in a processor and mix through sea salt.

The best way to reduce household waste is putting less on your plate. You can always go back for more. With leftover fruit, dried out skins blitzed into a powder add an intense orange or lemon zest to cakes.

It's better to run out of things than have food left over. Plan what you are going to eat every week.

Rice

Day old cooked rice is the best for fried rice, so if you're cooking jasmine rice, store any you don't finish in an air-tight container and pop it in the fridge straight away. It's recommended that you eat cooked rice within 24 hours. Keep it in the fridge and cook it from cold so it doesn't go clumpy when stir-frying it. We add spring greens, cherry tomatoes and a bit of soy sauce. You could add eggs if you wanted, too.

Lemons

Once squeezed, good unwaxed lemons can be finely sliced, massaged with salt, packed in a jar and after four days at room temperature, ferment into a lovely sour condiment. Use that as a relish or in any dish to replace lemons. We made pasta with some preserved lemon the other day. It was the best pasta I've had in a long time.

Bulk Shop

Why?

1. You'll save money.

Buying in bulk can save a family up to £500 per year. It may also mean spending less time in the supermarket, which can result in even more savings, as most of us often can't resist snapping up spontaneous purchases. If you're not there, you won't succumb to those little purchases that can quickly add up! Finally, shopping strategically means visiting the store less and maybe driving less — saving petrol/diesel/electricity and even more money.

2. You'll reduce the amount of food additives you consume.

Heavily processed foods, like many of the packaged convenience foods you find on the shelves of supermarkets, are stripped of a lot of their nutritional value. They also have a lot of synthetic additives and preservatives in them (not to mention all the extra sugars and salt). Buying bulk whole foods, opting for healthy snacks, and making meals from scratch is a far healthier way to live. A lot of foods in bulk shops only contain one ingredient!

3. You'll reduce food packaging (and chemical exposure!)

Buying in bulk uses less packaging. According to the Environmental Protection Agency, around 45 percent of waste in landfills is food packaging and containers. Some food packaging can release its chemical components into your food, too. If you're putting bulk beans into your own reusable bag, you won't have to worry about that!

4. You'll reuse.

Speaking of reusables, take your newfound bulk love a step further by using Bring Your Own durable containers instead of disposable bags at the store. Use glass or stainless steel containers. Reusable produce bags are also great. Just measure your containers before you fill them up so you won't be charged extra. If you must use plastic bags, try to reuse them several times and then return them to the store for recycling.

Prepare in bulk. If you prepare dishes on a larger scale then freeze portions then leftovers won't even exist. Not everything freezes well, such as rice, but you can cook those items fresh and be precise. Saucy things such as soup and curries can be batch-frozen ready to pull out whenever it's convenient.

Shop With Reusable Containers And Bags

Before going shopping, figure out what you need for the week and organise it so that you don't get caught out. Here are some suggestions:

- Glass jars for bulk items like flour, seeds, nuts, spices, tea, honey, coconut oil etc
- Metal containers or bento box for meat and fish
- Reusable shopping bags and reusable produce bags for produce and bulk foods like pasta or cat food

Different shops deal with containers in different ways. Some set scales out for you to weigh the empty containers. At other shops, customer service will weigh the jars for you. Then at some other shops, your request will completely baffle the staff... Keep going, though, and remember if that happens you're just first to be cool enough to bring re-usables, it's an amazing achievement!

Buy Fewer Ingredients

If you run out of something, you might find an alternative already in your kitchen cupboards. We buy lots of baking soda. We use it for baking, washing pots and pans, cleaning and making deodorant, it's good to buy things that have multiple uses.

Grow your own Food

Growing your own food can save you money as well as being an enjoyable hobby. If you don't have a big enough garden maybe you could consider getting an allotment. Maybe make it a family thing?

Or if the thought of starting an allotment is overwhelming you could ask your friends to do it with you and share the responsibility.

Think Twice Before You Buy Another Gadget

When our microwave broke we didn't bother getting a new one. My extended family thinks I'm strange because of that, but not only would it be a waste because we hardly used it anyway, it's bad for the environment and for our health. We don't have room for lots of appliances in our kitchen anyway because we have a "tiny home". Plus, we probably don't need what we don't already have, otherwise we would've got it already, right? Say no to shiny object syndrome :)

Ditch The Disposables

Use cloths instead of buying paper towels, we've repurposed old t-shirts to use as cloths. Any t-shirts that get ripped beyond repair goes into our drawer to be used as a towel. After all, it's just social conditioning that dictates using towels instead of old t-shirts, because they're usually the same thing. I do get strange looks from guests when

I'm stood in my kitchen and I take out a vest and start ripping it up though, until they see that there is a method to my madness :) My sister wonders how I run a kitchen without paper towels or cling film (plastic wrap). I've got two adventurous kids, so we've got enough ripped t-shirts to make rags that will last me the rest of my life. For cling film we actually use an eco friendly reusable and toxin free alternative to cling film that we also sell at HonestMiracle.com, our eco organic shop. Drink in, from a mug, instead of taking away, or invest in a reusable cup if you do take your coffee away.

Choose Biodegradable Sponges

The majority of sponges are made from oil-based plastics that can take up to 52,000 years to decompose in the landfill! Also, when left wet, sponges become a breeding ground for harmful bacteria, including E. coli and Salmonella.

Sponges made from 100% plant-based fibres help save the planet and lowers the risk of bacterial growth. Made from biodegradable materials, these sponges will break down easily in the landfill and naturally resist unwanted bacteria.

Recycle As A Last Resort

When plastics are recycled, they are actually downcycled—meaning even when reincarnated as toothbrushes, shopping bags or more plastic bottles, the plastic ends up in landfill eventually, unlike glass or metal, which can be recycled over and over without any degradation in quality.

Have you heard of America Recycles Day? It's a day in America designated to encourage us to put plastic water bottles into the recycling bin. Who dreamed it up? Big corps. So big corporations make a big mess and encourage us to clean it up, which doesn't actually work. BPA free isn't actually that great either, by the way. It's much better to cut off the rubbish at its source and refuse all this plastic junk.

How To Compost

A Traditional Compost Heap

Traditional compost heaps can be as huge as a three-part system or as simple as an open heap. The rules are the same in any system - add a good mix of greens and browns, sit back and wait!

A Wormery

Wormeries are perfect if you're short of space. They will work on balconies, outside the back door or even indoors under the sink. These will give you fantastic fertiliser for your houseplants.

A Bokashi Bin

A bokashi bin sits on your kitchen worktops. It will even degrade cooked scraps and small fish or meat bones. You add a special bran that activates the fermentation process and in two weeks you pop it on a compost heap.

Smart Cara

A Smart Cara is a fabulous solution if you don't have any outdoor space. It works like a dehydrator, reducing your food waste to 10% of its original size. You can then use this as a soil enhancer for potted plants.

Council Collections

Council collections are good if you don't have the space, time, or are unable to compost. Local councils will compost your food waste and turn it into soil enhancer, which you can often buy back from them, if you wanted to.

Quick Eco Kitchen Tips:

- As much as possible, put the lids on your pans! This keeps the heat in and helps shorten cooking times.
- When preparing vegetables, potatoes, pasta or other foods you'd cook in a pan, only use just enough water to cover. Using any more is wasted water and energy to heat the surplus.

- Use a slow cooker. As well as producing tasty meals and being incredibly easy to use, slow cookers use a lot less energy than your oven. Put your main ingredients in your slow cooker before work, when you get home you can add any extras. By dinner time you'll have made a delicious meal with minimal effort and much less energy consumption.
- Choose the right size burner for your pan. Too big and you'll waste heat and energy, too small and it will take longer to cook your food.
- Only run the dishwasher when it's full. If you're not doing this, you are probably wasting a lot of energy. Remember: a by-product of energy creation is greenhouse gas! Fill the dishwasher up during the day, then run a cycle in the evening before bed, I'm always a bit wary of running the dishwasher overnight, I know some people do it but I've seen what devastating effects leaving on big appliances like dishwasher or dryers overnight can have, so I don't recommend that.
- Hand wash dishes with the plug in the sink. If you need a few items before running your dishwasher, or you don't have one, this tip is a must! Washing dishes under running water is a huge waste, especially if you're using hot water.
- Adjust the thermostat in your fridge. The ideal temperature is between 1.7 to 3.3 degrees celsius, and for your freezer between -18 degrees celsius. Any lower and you could increase its energy consumption by up to 25%!
- Use an ice tray. Fridge ice-makers are massive energy drains, increasing consumption by up to 20%! Using an ice tray is simple, easy, and a lot more economic.
- Purchase energy efficient appliances. This one is for the long term, and something to consider when replacing an appliance. Almost all models are now at least 'A' rated. This doesn't mean that they are all the same. Models can be A+, A++ rated or even A+++ which is quite a big leap in energy efficiency. The greener the arrow for the fridge or freezer, the lower its energy consumption.

HUMMUS

It's pretty impossible to get shop bought hummus that's not in plastic, so I'll start off with this delicious recipe.

Ingredients

1/2 cup dried chickpeas, soaked and cooked
2 tablespoons tahini
1 tablespoon lemon juice
1/4 teaspoon sea salt
1 clove peeled garlic
1/4 cup olive oil

Instructions

1. Cook beans in a pot on the oven, in a slow cooker or in a pressure cooker according to instructions.
2. Drain beans and reserve some of the liquid.
3. Add the chickpeas, tahini, lemon juice, salt and garlic to a food processor. Process until smooth.
4. While the food processor is running, slowly pour in the olive oil. Process until all the oil is completely mixed in.
5. Makes about 1 1/2 cups.

ROASTED CHICKPEAS

Ingredients

2 cups dried chickpeas (makes about 5 cups when cooked)

2 teaspoons cumin

1 teaspoon oregano

pinch of cayenne pepper or to taste

3/4 teaspoon salt

1/4 cup olive oil

Instructions

1. Soak chickpeas for at least six hours. Cook in a pot, a slow cooker or a pressure cooker.

2. Drain, rinse, pat dry and place cooked chickpeas in a medium-size bowl.

3. Combine olive oil, salt and spices in a small bowl.

4. Pour olive oil mixture over chickpeas and toss.

5. Dump chickpeas out onto a rimmed baking sheet. Give the sheet a few shakes to spread out the chickpeas evenly in a single layer.

6. Roast at 200 degrees Celsius or gas mark 6 for 30 minutes. Stir every 10 minutes.

7. Chickpeas are ready when they have turned a rich, golden brown colour.

CAULIFLOWER COUSCOUS

Ingredients

2 tbsp olive oil

1 diced onion

1 clove minced garlic

1 head grated cauliflower

1/4 tsp cinnamon

3/4 tsp salt

1/2 cup pine nuts

1/2 cup raisins

2 tbsp minced cilantro

Juice of 1/2 lemon or to taste

Pepper to taste

Instructions

1. Grate the cauliflower with a cheese grater or in a food processor, using the blade attachment.

2. Over medium heat, saute the onion and cook for about 5 minutes or until softened. Add the garlic. Cook and stir for about a minute. Add the cauliflower, cinnamon and salt. Cook and stir for 10-15 minutes or until tender.

3. Stir in the pine nuts, raisins, cilantro, pepper and lemon juice. Stir for about a minute. Serve.

PUMPKIN DAHL

Ingredients

1 cup red lentils

3 cups water

2 tablespoons oil for sautéing

1 tablespoon cumin seeds

1 chopped onion

4 cloves minced garlic

1 serrano or small jalapeño pepper or 1 tablespoon fermented hot peppers, minced

1 teaspoon turmeric

1/2 teaspoon coriander

1 teaspoon fenugreek seeds

1 teaspoon dry mustard seeds

1 pound medium-size tomatoes, chopped (about 2 cups chopped into large chunks)

I cup pumpkin purée (make the purée in the oven or in a pressure cooker)

Juice of 1 lime or lemon or 1/4 cup preserved lemon juice

1 teaspoon salt or to taste

A few tablespoons of chopped cilantro

Instructions

1. Rinse lentils and add to a medium-size saucepan along with the water. Bring to a boil, turn down the heat and simmer for about 10 minutes or until tender. They will look soupy once cooked.

2. Heat oil in a large saucepan over medium-high heat. Add the cumin seeds and cook until fragrant, about one minute.

3. Add onions, garlic and fresh jalapeño. (Wait to add fermented jalapeño later.) Sauté over medium heat for five to ten minutes until the onions are translucent.

4. Add spices and cook for a minute.

5. Add tomatoes and cook for about five more minutes.

6. Stir in pumpkin purée, cooked lentils and their liquid, fermented jalapeños if using, lemon or lime or preserved lemon juice and salt. If the consistency remains thick, add a bit of water. Heat through. Garnish with fresh cilantro.

7. Serve with rice, naan or roti.

PESTO SAUCE

Ingredients

1 bunch basil leaves (makes about 120 grams or 2 cups packed)

2/3 cup pine nuts or walnuts

2 cloves garlic, crushed

3 tbsp nutritional yeast

1/2 tsp salt

2/3 cup olive oil

Instructions

1. Put basil, walnuts, garlic, nutritional yeast and salt in the food processor. Pulse until well combined.
2. With food processor running, slowly pour in olive oil. Process until smooth. Store in a glass container in the fridge.

CHANA MASALA

Ingredients

1 1/4 cups dried chickpeas

2 tablespoons coconut oil (or other oil)

1 teaspoon cumin seeds

1 medium onion diced

4 teaspoons peeled, minced ginger

4 garlic cloves minced

1 to 2 serrano chiles or jalapeños, minced

2 cups roasted tomatoes

2 teaspoons garam masala

1 teaspoon ground coriander

1/2 teaspoon turmeric

1 teaspoon salt or to taste

1/2 cup water

Instructions

1. Soak chickpeas for six hours or longer and cook in a pot on the stove, in a slow cooker or in a pressure cooker until done. I cook chickpeas in 5 minutes in my pressure cooker (I love it!).

2. In a large skillet or Dutch oven, heat the oil over medium heat. Add the cumin seeds and cook, stirring until fragrant, about 1 minute.

3. Add the onion, ginger, garlic, chiles and salt. Cook, stirring occasionally, until the onions have softened, about 5 minutes.

4. Add the garam masala, coriander, turmeric and salt and stir to coat the onion mixture. Cook, stirring occasionally, until fragrant, about 1 minute.

5. Stir in the tomatoes. Break them up with the back of a wooden spoon.

6. Stir in the chickpeas and the water. Bring to a simmer. Reduce the heat to medium low and simmer uncovered, stirring occasionally, until the sauce has thickened slightly, about 20 minutes. Serve with rice or naan.

BANANA PANCAKES

Ingredients

1 over ripe banana, I keep mine in the freezer
1 3/4 cups white spelt flour all purpose flour is also okay
1 1/2 Tablespoons sugar
1 1/2 Tablespoons baking powder
3 Tablespoons organic oil of your choice - coconut oil works well
1/8 teaspoon salt
1 teaspoon vanilla
1 cup non-dairy milk
vegan buttery spread for the pan

Instructions

1. Mash the banana in a large bowl. Add the canola oil, vanilla, and non-dairy milk. Stir to combine.
2. Add the white spelt flour, sugar, baking powder, and salt. Mix together.
3. Place a tablespoon of vegan buttery spread in a large skillet, and heat over medium low heat.
4. Once the pan is hot, spoon the batter into the pan. Cook the pancakes low and slow, it may take about 3-4 minutes on each side. Once the pancakes are done around the edges, flip them over with a thin spatula. Cook for about 3 or 4 minutes more, then remove from pan.
5. Repeat with remaining pancakes. Top with syrup.

LEFTOVER VEGETABLE PIE

The best thing about a leftover pie is that anything goes, really! Just pop all leftover vegetables in, although this recipe is really delicious.

Ingredients

1 kg maris piper potatoes

Olive oil

200 g vegan cheese, grated

1 large garlic clove crushed
1 red onion, finely chopped
3 carrots grated
1 broccoli
3 celery stalks finely chopped or grated
1 fennel root roughly chopped
1 red chilli finely chopped
1 400g can chickpeas drained and rinsed
handful fresh parsley finely chopped
1 lemon

Instructions

1. Pre-heat your oven to 200C
2. Slice the potatoes into small cubes, leaving the skin on. Place in a pot with boiling water for 12 minutes until soft
3. Meanwhile prepare the veg. Keeping the skin on the carrots, grate them straight into the baking dish. Add the chopped celery, fennel root, chilli (making sure to not waste any edible parts) parsley leaves and stalks, red onion, garlic, chickpeas and 100g of vegan cheese
4. To reduce food waste, chop the florets off the broccoli head and then grate the stalk. Add both to the baking dish along with the zest of 1/2 a lemon and the juice of the full lemon. Season with salt and pepper and mix all the ingredients

5. Once the potatoes are boiled, mash with olive oil (or how you prefer to do it) and add in the leftover 100g of vegan cheese and mash until smooth. Simply layer on top of the pie mix and place in the oven at 200C for 35 minutes

LEFTOVER VEGETABLE STEW

Ingredients

3 large garlic cloves crushed

1 red onion diced
350 g mushrooms chopped
1/4 cup tamari
4 carrots chopped
3 celery stalks chopped
500 g baby potatoes sliced into halves
1 can green lentils drained and rinsed
1 stem of rosemary finely chopped
3 bay leaves
2 400g tins tomatoes
300 ml veg stock
2 tbsp cacao powder

Instructions

1. Heat a large pot with olive oil. Add in the garlic, onion, mushrooms and Tamari. Stir and cook down for 5 minutes
2. Now add the carrots, celery, potatoes, lentils, rosemary, bay leaves, tomatoes, veg stock and stir
3. Add in the cacao powder, stir again until well mixed
4. Place the lid on the pot and simmer for 30 minutes
5. Serve with brown rice or quinoa and top with fresh parsley (optional)

SWEET POTATO AND BLACK BEAN STEW

Ingredients

2 tbsp olive oil/coconut oil
2 large garlic cloves, crushed
1/2 large red onion, diced
1/2 tsp ground coriander
1 heaped tsp turmeric
1/2 tsp cayenne pepper
1 pinch chilli flakes
1/2 tsp paprika
2 large sweet potatoes(roughly 800g), peeled and chopped into small cubes
1 tbsp tomato puree
2x 400g cans tomatoes
2x 400g cans black beans, drained and rinsed
400 ml veg stock
2-3 handfuls fresh spinach or kale
salt & pepper

Instructions

1. Firstly peel and chop the sweet potatoes into small cubes and dice the red onion
2. Heat a large pot on a medium heat with olive oil, garlic, onion with a pinch of salt & pepper and fry for a couple minutes
3. Add all the spices and mix together before adding the sweet potato along with a pinch of salt and pepper. Fry together for a couple minutes making sure to stir
4. Add the tinned tomatoes, tomato puree, black beans, veg stock and a pinch of pepper and mix together
5. Bring to a boil then reduce to a low heat placing the lid on the pan for 20-25 minutes (until your desired consistency, I like mine fairly thick)
6. Add the spinach and stir through, allowing it to wilt before taking off the heat

SWEET POTATO AND CHICKPEA CURRY

Ingredients

2 tbsp olive oil/coconut oil
2 garlic, crushed
1 pinch chilli flakes
2 tsp turmeric
1/2 tsp cayenne pepper
1/2 tsp paprika
2 tbsp tomato puree
2 large sweet potatoes, peeled & chopped
2 cans of coconut milk
2 cans tinned tomatoes
1 can chickpeas, rinsed & drained
200 g baby spinach leaves
Handful of coriander
1 lime
Salt & pepper
Instructions

1. First off peel and cut the sweet potato into small chunks (roughly 2cm cubes)
2. Heat a large pot with olive oil, adding the crushed garlic cloves
3. Once the garlic is sizzling proceed to add the tomato puree, all the spices and a pinch of salt and pepper into the pot and stir for a minute or so
4. Add the tinned tomatoes, coconut milk, sweet potato and a pinch of salt and pepper and mix well
5. Bring this to the boil making sure to stir and then place in a preheated oven at 200 degrees for 45 minutes (I leave the lid off the pan which makes the curry slightly thicker and creates a nice baked flavour) making sure to stir every 15-20 mins. You could just simmer on your hob if you wish.
6. After 45 minutes add the drained chickpeas and spinach and mix. Place back in the oven for 15 minutes

LEFTOVERS QUICHE

The Pastry

Ingredients

- 200 g gluten free flour
- 25 g chickpea flour
- 1 tsp Xanthan Gum (available in some supermarkets and health food stores)
- ¼ tsp salt
- 50 g stork block (available in most supermarkets)
- 50 g Trex (available in most supermarkets)
- 4 tbsp water (added slowly)

Instructions

1. To a large bowl sieve in both flours to remove any lumps then add in the Xanthan gum and salt and mix
2. Now add in the stork and trex and with rub with your fingertips until it looks like breadcrumbs
3. Add in the water one tablespoon at a time until the mix comes together. It should be a little sticky and not dry!
4. Roll out between 2 sheets of cling film into a round shape to roughly 3mm thick (see picture above)
5. Carefully place into a 9 inch quiche/tart tin and press around the edges

The Quiche

Ingredients

350 g silken tofu
4 tbsp nutritional yeast
¼ cup gram flour
¼ tsp garlic powder
1 tsp dried basil

1 tsp dried oregano
2 handfuls spinach
1 handful fresh basil
Large Pinch salt & pepper
200 g fresh tomatoes (optional)
A handful of sundried tomatoes (optional)
1 red onion, sliced (optional)

Instructions

1. Pre-heat your oven to 180C
2. To a food processor add the silken tofu and blend
3. Now add the gram flour, nutritional yeast, herbs, garlic powder, salt and pepper and blend until smooth (you may need to press down the edges)
4. Whilst this is blending simply heat the spinach in a pot until it wilts
5. Now to a large mixing bowl add the tofu mixture and spinach along with fresh basil, chopped sundried tomatoes and red onion and mix
6. Pour the mix directly into the pastry tin and top with the sliced tomatoes
7. Bake in the oven for 30 minutes at 180C
8. Remove from the oven, allow to rest for 5 minutes before carefully removing from the tin

PEANUT BUTTER AND CHICKPEA CURRY

Ingredients

2 tbsp olive oil
3 garlic cloves crushed
2 small red onions diced
2 400g tinned chickpeas drained
4 tbsp peanut butter
1 tsp paprika
1/2 tsp cayenne pepper
1 Pinch chilli flakes
1/2 tsp ground coriander
1 tsp veg stock powder
1 Can coconut milk
1 Can chopped tomatoes
2 large handfuls spinach
1 handful Jumbo peanuts unsalted
Salt and pepper

Instructions

1. Heat a pan on a medium heat with the olive oil and add the crushed garlic, onions and a pinch of salt & pepper. Cook down for a couple minutes
2. Add all the spices, peanut butter, chickpeas, salt and pepper. Stir well and fry together for a couple minutes
3. Add the tinned tomatoes, coconut milk and the veg stock powder and mix well
4. Bring to a boil and then reduce to a low heat for 15 minutes (lid off) making sure to give it a stir every 5 minutes
5. Stir in the spinach and cook for a further 5 minutes (or to your desired consistency)
6. Dry fry a handful of the jumbo peanuts until lightly browned to serve on top.

ROCKY ROAD

Ingredients

70 g pecans
70 g almonds
50 g pistachios (unsalted)
50 g dried cranberries
40 g cacao powder (roughly 1/2 cup)
3 tbsp maple syrup
2 tbsp almond milk
200 g coconut oil melted

Instructions

1. Firstly add all the nuts to a pan and lightly toast for 5 minutes
2. Meanwhile make the chocolate sauce by adding the melted coconut oil, cacao powder, almond milk and maple syrup to a cup blender and blend until smooth (30 seconds)
3. Line a small baking tray (9x6") with baking paper and place in half the nuts and cranberries. Pour over the chocolate sauce then top the rest of the nuts and cranberries on top
4. Place flat in the freezer for 15-20 minutes. Remove and slice into squares or bars

CHOCOLATE CHEESECAKE SQUARES

Ingredients For Base:

200 g almonds
100 g pitted dates
2 tbsp water

Ingredients For The Cheesecake:

200 g cashews soaked for 30 minutes in hot water
100 g coconut fat top part in a can of coconut milk roughly 6 tbsp
9 tbsp coconut water from the bottom of the can of coconut milk
3 tbsp maple syrup
1/2 Juice of lemon

Ingredients For Chocolate Topping:

6 tbsp coconut oil melted
3 tbsp maple syrup
40 g cacao powder

Instructions

1. I used a 9 inch square cheesecake tin, you can try with others if you wish for a different shape
2. Soak the cashews for 30 minutes in hot water
3. First make the base by adding the almonds and dates into the food processor and blend into crumbs, then add the water and blitz again to make it sticky
4. Place a sheet of baking paper inside the tin then place the nut mix into the bottom and press down until evenly spread
5. Drain the soaked cashews and add to a high powered processor or a large cup blender along with the lemon juice, maple syrup, coconut fat, coconut water and blend until very smooth (this can take up to 5 minutes, make sure to press the leftover bits on the side down to make the whole mix smooth)
6. Now Pour the mixture on top of the base making sure to smooth over the top
7. Add to the freezer for 2-3 hours (you are looking for the cheesecake layer to be firm)

8. Just before the 2-3 hours is up make the chocolate sauce topping by adding the melted coconut oil and maple syrup to a Bain Marie and slowly stir in the cacao powder
9. Remove the cheesecake from the freezer, gently pour over the chocolate mix and place back in the freezer for 30 minutes
10. Simply slice it up and they're ready to serve

GOOEY CHOCO LAVA POT

Ingredients

3 tbsp oat flour (make your own by simply blending GF oats into flour, takes 30 seconds)
3 tbsp raw cacao powder
1/2 tsp baking powder
1 1/2 tbsp maple syrup
60 ml unsweetened almond milk
Spoonful nut butter optional
Pinch of salt

Instructions

1. Pre-heat your oven to 200C (make sure it's up to temperature before cooking the choc pot)
2. Create your oat flour by adding gluten-free oats to a blender for 30-60 seconds until it forms a flour
3. Now to a mixing bowl add the oat flour, cacao powder, baking powder, tiny pinch of salt and whisk until well mixed with no lumps
4. Now add the maple syrup and pour in the milk slowly whilst hand whisking until smooth
5. Add a spoonful of nut butter to your pot (optional)
6. Pour the mixture into your pot/mug and place in the oven for 10-12 mins or microwave for 40-80 seconds (the time for the oven really depends on your oven and can take a couple tries to get it right)
7. Let cool for a minute before eating and it should still be gooey in the middle and slightly firm on the top and sides

ZERO WASTE BANANA ICE CREAM

One of the best things about bananas is that when frozen, its texture is very similar to ice cream! Yum!

I have always been meaning to try freezing smaller pieces and then blitzing in the food processor, but I usually like to keep things simple.

This is healthy enough to have as breakfast, or a snack anytime and of course it is suitable for any diet as long as there's no banana allergy.

It can also be an occasional treat for your dog (if you happen to drop one on the floor!)

Instructions

1. Grab a banana
2. Peel it (compost the peel if you can)
3. Break the banana in half, or cut length-ways
4. Place on a baking tray
5. Freeze it!
6. Wait at least an hour and start eating your ice-cream!

It's best to leave it overnight, but if you just can't wait, an hour or two is fine. For toddlers, you can slice the banana length-ways into thinner strips, so it is a little easier to hold.

If you want to jazz it up a bit, you can make up your own variation, here are a few suggestions:

Drizzle dark chocolate on top
Poke several dark chocolate chips into the bananas

Dip in dark chocolate and roll in almond pieces

Coat in coconut yoghurt and add sprinkles

SNACKS

ROASTED PUMPKIN SEEDS

Pop the seeds from your pumpkins in olive oil to coat, along with salt and, if desired, spices (cayenne pepper, cumin etc.). Spread in a single layer on baking paper or a glass tray. Roast at 160 degrees celsius in a fan oven or 180 degrees in a conventional oven for 20 or 30 minutes, stirring every 10 minutes until golden and crunchy.

SAUCEPAN POPCORN

Use a medium-size saucepan. Add 1 teaspoon of salt, 1/2 cup popcorn kernels and two tablespoons of coconut oil or olive oil, or a combo of the two. Put the lid on the pot, turn the cooker to high heat and shake the pot continuously. Once the kernels begin to pop, they will pop quickly, in about a minute.

GRANOLA BARS

Ingredients

1/2 cup unsalted peanut butter

1/3 maple syrup or brown rice syrup

1 egg replacement: combine 2 tablespoons water, 1 teaspoon oil and 2

teaspoons baking powder

2 tablespoons coconut oil

1 1/2 teaspoon vanilla extract

3 cups old fashioned rolled oats

1/2 cup seeds of whatever type you have on hand (C used pumpkin and flax seeds)

1/2 cup dried cranberries

1/2 cup packed light brown sugar

1/2 teaspoons salt

1/3 cup bittersweet chocolate chips or pieces

2 tablespoons cocoa powder

Sunflower seeds to press into top

Instructions

1. With a mixer, beat together wet ingredients: peanut butter, honey or maple syrup, egg replacement, coconut oil, vanilla extract.

2. Combine dry ingredients in a separate bowl.

3. Add dry ingredients to wet ingredients.

4. Put mixture into a greased 13 x 9" baking dish. Press down into pan until flat and relatively even. Sprinkle with ample sunflower seeds and press firmly into the granola mixture.

5. Bake at 160 degrees for a fan oven or 180 degrees for a conventional oven for about 15 minutes or until golden.

6. Allow to cool. Cut into bars. Store in a glass jar or container.

BANANA BARS

Ingredients

6 ripe bananas
400 g gluten free oats
1 tsp cinnamon
2 tbsp water

Instructions

1. Pre-heat your oven to 200C
2. To a large mixing bowl add the peeled bananas and mash with a fork
3. Now add in the oats, cinnamon and water and mix together
4. Line a square baking tin with baking paper and pour in the mix. Press down firmly with the back of a fork to compact it into the tin
5. Bake in the oven for 20-25 minutes at 200C.
6. Remove from the oven, allow to cool for 10 minutes before slicing into bars

Store in a sealed container and consume within a few days.

CRUNCHY PROTEIN BALLS

Ingredients

330 g crunchy peanut butter (roughly 9 heaped tbsp)
90 g ground almonds
7 tbsp maple syrup
1 pinch salt

Instructions

1. In a large mixing bowl add the peanut butter, ground almonds, maple syrup and salt. Whisk until it's well mixed together
2. With a spoon, scoop up a small amount and roll into a rough ball, no need to be perfect as we will roll them again later which will make them smooth
3. Place each ball on a tray and place in the freezer for 30 minutes
4. Take the balls outs the freezer and roll again in your hands until smooth
5. They are now ready to enjoy

"BOUNTY" BARS

Ingredients for inside:

1/2 cup desiccated coconut (50g)

1/2 cup coconut flour (50g)

3 tbsp maple syrup

2 tbsp coconut oil, melted

Ingredients for outside:

4 tbsp cacao powder (25g)

3 tbsp coconut oil, melted

2 tbsp maple syrup

Instructions

1. Add the desiccated coconut and coconut flour into a large mixing bowl and hand whisk, making sure there are no lumps
2. Now add the maple syrup and melted coconut oil and mix well
3. In your hands, form bar shapes with the mix (this doesn't need to be perfect, I prefer mine more rustic with random shapes)
4. Now place onto a plate and put them into the fridge for 10-15 mins to harden
5. When there is 5 minutes left on the timer, add the melted coconut oil and maple syrup over a bain-marie and stir
6. Now add the cacao powder and mix in well making sure there are no lumps
7. Remove the coconut bars from the fridge and simply dunk into the chocolate mixture, making sure to cover evenly
8. Place the bars onto a cooling rack or baking paper and place back into the fridge for 1 hour to harden (you can add to the freezer if you want to speed up the process)
9. Just before serving I like to drizzle the leftover chocolate mixture over the top along with some desiccated coconut

CHOCOLATE TRUFFLES

Ingredients

6 tbsp coconut oil melted

6 tbsp cacao powder

5 tbsp maple syrup

Pinch salt

Optional Toppings:

Cacao powder

Crushed pistachios

Cacao nibs

Instructions

Add the melted coconut oil and maple syrup to a Bain Marie and mix together

Slowly add each tbsp of cacao powder whisking until you have a smooth mix

Add a pinch of salt and mix in

Transfer to a bowl and place in the freezer for 45-60 minutes

Remove from the freezer (the mix should be fairly firm) and scoop out with a spoon into your hands and roll into balls

Choose a topping to cover it in (ground pistachios, cacao nibs or cacao powder work well)

Place back in the freezer for 15-30 mins

Bathroom

Bathroom products are large offenders for excess packaging, which as we know are very likely to end up in landfill. Face creams, shampoos, toothpaste and countless other products all come in plastic as the norm. This packaging is often not even being recycled! The most common bottles and packaging not known to be recyclable are bleach, shampoo, conditioner, bathroom and kitchen cleaners and soap dispensers.

If you need to use plastic, choose recyclable plastic, and pop a reusable bag on the back of your bathroom door for your empty plastic bottles (and toilet roll tubes) so they can easily be put in the main recycling bin.

Even more interestingly, the products used to clean your bathroom are also unlikely contributors to climate change. Aside from plastic packaging, the process required to make synthetic scents actually uses masses of energy and thus contributes to climate change. In fact, the chemical sector is the third largest source of industrial CO2!

This goes to show that being zero waste cuts down on more waste than we can actually see. So your rubbish bins will be emptier and industrial waste and greenhouse gas production could also drop!

Plastic comes in all forms in the bathroom. Start by looking at what you use every day. Do you use cotton buds, cotton wool, floss, disposable razors and plastic toothbrushes? They're all single-use plastic, after its very short life, it heads straight in the bin and will most probably end up in landfill. We can't go without personal hygiene items, instead, we can find some more sustainable and zero-waste alternatives.

Use A Bamboo Toothbrush

As the focus of environment talks shifts to single-use plastics, more people are looking for ways to swap everyday items for something more eco-friendly. Bamboo toothbrushes have become more and more popular in recent months, and are a safe and hygienic way to brush your teeth. An easy swap to make.

Use Soap Bars

Instead of buying soap in packaging, go for bars of soap, preferably toxin free for your health and for our water health. A Swiss study found that liquid soap has a 25% larger carbon footprint than soap bars, so you're reducing waste and fighting climate change!

Shampoo and Conditioner Bars

So much waste is created by shampoo and conditioner bottles and most of these bottled potions are actually bad for our hair, scalp and health. Switch to shampoo and conditioner bars for better health, skin and much less waste.

Conserve Water

According to the National Trust, the average household in the UK uses 330 litres of water each day. That's shocking isn't it?

A great way to conserve water is by installing low flow faucets or low-flow aerators on your taps. It's pretty simple. Unscrew the old aerator, apply some pipe tape on the new one and screw it on the faucet. And that's it :)

Most people have heard to switch off the water when brushing your teeth, but we can also switch it off when we're in the shower washing our hair, or shaving, every little drop counts.

Wash Your Hair Less Frequently

How often do you wash your hair? If you wash it every day, you could be causing unnecessary damage, while also wasting water. Learn how to keep your hair fresh between washes to ensure you don't need to reach for the shampoo every day.

Install Low Flow Shower Heads

Low flow shower heads use less water every time you shower which in turn saves you energy and reduces your carbon footprint even further.

Some local councils sell low flow shower heads at a reduced price so definitely worth doing a bit of research before you buy one. You will need to install it but they are fairly easy to set up.

Fix Leaky Taps and Showers

Do you often hear a familiar dripping coming from your bathroom? If you've got something leaking, get it sorted as quick as you can. Whilst the odd drip here and there might not seem like a lot, it can soon add up if left unresolved. You might need new taps installed to fix the problem, and you can find some here. Remember to check your toilet too to make sure it's not dripping unnecessarily.

Take Less Baths

Baths are a great way to relax and unwind, but they can use up a lot of water. Limit your baths and possibly share bath water if someone else in the family wants to take one. Keep the water level to where it needs to be and don't overfill the bath.

Use a wooden Toilet Brush

Or bonus points if it's made from bamboo!

Ditch The Cotton Buds

Cotton buds are one of the biggest offenders of single use plastic waste. Many people flush them down the toilet, meaning they end up in the water where birds, fish, and animals can easily swallow them. We need to be more mindful of what we flush down the toilet to stop dangerous waste from entering our oceans. There are some great alternatives to cotton buds you can use, such as a reusable ear cleaner, or if you don't want to use a reusable solution you can go for biodegradable cotton buds such as bamboo ones.

Invest In A Dual Flush Toilet

If you're serious about saving the environment, one thing you can do is invest in a dual flush toilet. These modern toilets use only half the water flow of a regular toilet, saving tonnes of water each year.

You can buy parts to turn your own toilet into a dual flush toilet to help you be more eco-friendly every day

Ditch The Toilet Paper!

Now this one is pretty hardcore even by my standards but you'd be amazed by how many people out there are choosing to ditch toilet paper. The two most popular alternatives are reusable toilet paper (yes you read that right) and installing bidet toilets or attachments.

For me personally, I think reusable toilet paper would be a step too far and I'm pretty sure the other half wouldn't be too happy about it either. I am considering having bidet attachments installed to our toilets though.

Buy natural toiletries and beauty products

Your toiletries and beauty products can contain a lot of chemicals and unnecessary additives. Buying products made from natural ingredients will mean that you help the environment, while also saving your skin from potential irritants.

Use Reusable Sanitary Products

One of my best decisions was moving on from disposable sanitary pads to lovely toxin free reusable pads. I stopped using tampons after giving birth, having kids made me rethink all toxins. I was never a fan of them anyway, but society pressures...

So, I use Reusable Organic Bamboo Sanitary Pads from Honest Miracle, they're honestly amazing. I'm not just blowing my own trumpet because they're our products, but they're so comfortable (many customers have said this too) and many have even experienced less pain since using our organic bamboo reusable pads.

Urinary Incontinence Pads

Just a note that some reusable pads can be used for urinary incontinence, too. A lot of our customers at Honest Miracle use them every day with great success. It's always amazing to get an email from a woman saying your product has changed their life. That's what I do it for. Random thought sharing there, sorry :)

Use Natural Deodorant

You can buy natural deodorants, but I also recommend trying coconut oil and bicarbonate of soda. Make it into a paste and rub it in. I swear by it! My Mum loves it too. Just beware that during the process of your body getting used to the lovely natural products and ridding your body of the toxins, your underarms could smell a bit more, but stick with it and when your body gets used to it, you won't go back! No toxins and no plastic deodorant containers!

Bedroom

Wardrobe

Did you know that there are an estimated £30 billion worth of unused clothes in our wardrobes?

Plus, according to research done by Matalan, the average woman spends ONE YEAR of her life agonising about what to wear! And with the amount of awareness on climate change these days, it's not just about matching outfits any more, we have to take sustainability into account as well.

Instagram users are buying fast fashion to wear just once for a #ootd (outfit of the day) photo, then throwing them out! The fashion industry is driven by trends and our desire to follow them. Because people's interest in clothes has grown, the number of trends and their frequency has also grown. The "need" to have the latest trends fuels a throw away culture where clothing is seen as useless after a few uses. This wasteful habit of dumping clothes after only a few uses sees clothes end up in landfill shortly after they are bought.

A study of 2,000 people found that one in 10 would throw away an item after being pictured online with it three times! The survey also revealed one in five throw their clothes in the bin rather than give them to charity.

And, according to Oxfam, the global textiles industry is devastating for people and our planet – contributing to 8% of the world's greenhouse gas emissions which affect climate change. **That's more polluting than aviation and shipping combined!**

On the production end, unsustainable manufacturing processes use huge amounts of water, chemicals and carbon to create clothes on an industrial scale. These clothes are called 'fast fashion', because they are made extraordinarily quickly and are designed to last for a short period of time before a new trend takes over. This fast and cheap method of production also produces a lot of textile waste!

The good news is there is a lot you can do to combat this!

Rent clothes for special occasions - There are plenty of clothes hiring websites to choose from.

Buy second hand - When you do need to buy, buy from a charity shop or second hand from the comfort of our own home on Facebook Marketplace, some people even deliver!

Clothes swapping

Another way of keeping your wardrobe zero waste is by swapping clothes instead of buying them. Swap with friends and family, or even host your own 'Swishing' party where every guest brings a few items and then they get to 'shop' from what others have brought. This is a fun way of reducing your fashion waste and adding a few new pieces to your wardrobe, whilst freeing up some space too.

Make sure you feel the love

If there is something that catches your eye on the high street, keep the tags on and if you don't wear it within a week, take it back. If I'm bought gifts I always make sure I try things on straight away and if I don't **feel** amazing wearing them, I'll take them back.

Do some research into the brand that you are buying from. Buying less but higher quality clothing can significantly reduce waste. There are also some brands that utilise stock fabrics left over at the end of the season, old denim or even abandoned fishing nets pulled from the ocean to make clothes.

Cotton is particularly important to research thoroughly as it's actually one of the most wasteful materials to produce. Choosing organic cotton has a huge environmental saving compared to conventional cotton. Look out for the GOTS symbol or the Soil Association's symbol on the label. Or try one of the more sustainable fabrics such as bamboo or hemp as an alternative to conventional cotton or polyester.

If you have clothes that are old or too worn out to continue wearing them, cut them up into cloths to use around the house or try making something out of the fabric. Re-using your old clothes will keep them out of landfill, save you money and save you a trip to your local donation bin.

If you can't find a use for them at home, then still think before you bin. There are places that collect "rags" to be shredded to make insulation material for the building industry.

You might find that a local charity shop has a rags collection, or sometimes a local school might collect them.

Anything we can do to keep fabric out of landfill is a good thing.

Also washing and storing clothes properly increases their longevity and means we don't have to throw them away unnecessarily.

Resources: QUIZ: Are your clothes damaging the planet? https://www.bbc.co.uk/news/science-environment-47156853

Reuse Silica Gel Packets

Don't throw them out! Silica gel sachets can be reused and re-purposed in so many ways and I bet you can find at least one useful way to use them around the house in this list.

Silica gel is often found in store bought products like handbags, shoe boxes, clothing, furniture, and there are many surprisingly simple ways to reuse them.

As you would expect, over time, they do lose their effectiveness, but you can refresh them by simply baking them in your oven for around 10-15 minutes at 150 Degrees Celsius. The moisture absorbed will dry out and the silica sachets will be as good as new.

Here's what you can reuse your silica sachets for in the bedroom:

- Keep shoes fresh and dry
- Prevent musty smells in clothing and storage drawers
- Pop a few in your jewellery box to prevent tarnishing
- Place in the bottom of handbags when not in use
- Store a few sachets in your gym bag to combat moisture and odours
- Useful for keeping seasonal clothes including winter slippers, hats, scarves and gloves fresher for longer

Silica Gel In The Bathroom

Bathrooms are one of the most common places to find humidity and condensation and it's unavoidable. You may not realise that this can be affecting the items in your bathroom cabinet. You will still need to use fans and crack open the windows from time to time, but give these a go too:

- Got a safety razor? Keep silica sachets with your razor blades to prevent rust.
- Place a few in your makeup bag to ensure powdered items stay fresh and unaffected.
- Keep your bath bombs and bath salts dry and maintain their fizziness longer.
- Place a few silica sachets underneath your mirror and on your windowsill to prevent fogging and mist.
- Stop your nail scissors, tweezers and eye liner sharpeners rusting or tarnishing.
- Place some in your bathroom cabinets which are prone to humidity and condensation.
- Keep a few sachets with items such as cotton buds, toilet paper, spare soaps and shampoo bars or whatever else you are stocking up on.
- Pop silica gel into your travel pill box to keep vitamins and medicine fresh.
- Use in travel toiletry bags to soak up any small spills.

Here are a fair few other uses for silica gel (I love reusing these!):

Laundry

- Store several in your towel cupboard to capture dampness.
- Throw a sachet into your washing powder container to prevent clumping (stick it to the lid if you're worried about it accidentally going into the washing machine).
- Keep silica in your laundry cupboards and stop musty odours.

Kitchen

- Store a few silica sachets in the base of your utensil holder, to absorb excess moisture on wood and bamboo utensils.
- Add into any kitchen drawers that is prone to heat changes, like next to the dishwasher.
- Store with dried herbs to maintain freshness (they can help you dry your own fresh herbs too).
- Prevent dry items like epsom salts and bicarb soda from clumping by sticking silica sachets to the lid so it doesn't get accidentally scooped up.
- Add one or two sachets to keep dishwashing powder from clumping or sticking to the inside of the lid.
- Stop homemade pet treats getting soggy.

- Keeps bread fresh and mould free, especially if you store it in plastic bags or containers.
- Store with homemade crackers or cookies to keep the crunch longer.
- Keep tea leaves and ground coffee fresher and moisture free.
- Keep some with your steak and cutting knives to prevent rust spots.

Electronics

- Keep any 1980's cassettes and videotapes in better condition.
- Add silica gels into your cable storage boxes to collect moisture.
- Pop one into your camera bag to prevent lens fogging or any humidity.
- Deal with mobile phone dampness.
- Keep them with your spare batteries.
- Storing one in your earphones case, especially great if you use them while exercising, the sachets can absorb extra moisture.
- Store silica with any electrical and battery operated devices that may be exposed to humidity and moisture.

Other random uses for silica gel sachets around the home:

- Remove those musty smells from old books
- Keep photos and photo albums moisture free
- Store in your seed collection containers to preserve them longer
- Useful for drying flowers and storing them moisture free
- Add a drop of essential oils and use as a scented freshener
- Pop a few inside your door draught stopper
- Insert into cushions, just remember to remove before washing
- Pop a few silica sachets into your display cabinets
- Store with board games
- Silica can help keep the blade on pencil sharpeners in good condition and prevent rust
- Medicine cabinets and first aid kits that are prone to temperature changes
- Behind picture frames to absorb any moisture that may damage your art
- Prevent rust and tarnish in your sewing box and needles
- Keep your yoga mat fresh and sweat free
- Pop one in with your matchboxes and candles to keep them dry
- Prevent condensation on inside windows anywhere in the house
- Place with important papers and documents to maintain their condition
- Keep kids art work from being damaged by moisture

- Any storage containers that may get moisture, especially if exposed to heat variations
- Christmas decorations like play dough and clay ornaments you want to keep dry
- Party decorations, especially paper and plastic that can become easily damage
- Protect wrapping paper, cards and tissue paper
- Store a few silica sachets inside suitcases and travel bags

Reusing Silica In Your Shed

- Store silica gel in tool boxes to prevent tools from rusting
- Keep a few with spare nails and screws in the shed
- Great for tents, camping gear and sporting equipment to prevent musty smells and condensation
- Inside gardening gloves that may get sweaty and smelly

Silica In Your Car

- Put on the dashboard and around windows to prevent interior fogging
- Add a few to the glove box to keep it smelling fresh
- Add a few drops of essential oil to make a DIY air freshener
- Store in key areas around your camper van, especially in the off seasons, to keep it free of musty odours

Please keep these away from children and pets who may attempt to eat them. They aren't poisonous, however, if consumed they can cause major issues since they absorb moisture and some do contain other ingredients. So find a safe storage place. When collecting silica gel sachets, be sure to save them up in an airtight jar and don't leave them to soak up moisture in the air.

Clean with care

Stay away from synthetic fragrances as we know these chemicals produce a lot of greenhouse gases and they're also toxic. Instead, try and use natural alternatives like vinegar and pop it into reusable spray bottles. By creating your own cleaning products, you avoid so much waste and don't have to worry about using harmful chemicals. Less waste means less worry!

Citrus Peel Infused Vinegar Natural Cleaner

This is great for any areas that get greasy and is a fab way to make use of citrus scraps.

What You'll Need:
Citrus Peels
Vinegar

Instructions

For this you can use lemon, lime, orange, mandarin, grapefruit or any other citrus peel from the fruits you eat. You can also combine them in one jar if you wish. My absolute favourite is tangerine and in less than 24 hours it gives off a beautiful smell.

Cut the peel into strips, add to your jar and top up with white vinegar. Secure the lid and shake.

Allow it to sit under your kitchen sink or in the pantry away from sunlight and heat sources.

After several days, the citrus will have infused and it is ready to use. For a stronger cleaner, allow it to sit for at least 2 weeks.

Pour the mix into a spray bottle using a funnel to avoid transferring the peels. This can be diluted with 30-50% cooled boiled water.

Here are top uses of citrus infused vinegar around the home:

1. Greasy, baked on messes on the Stovetop and oven door.
2. Spray baking dishes and griller trays to easily remove fat and grease.
3. Microwave Splatter and odours - Simply spray and leave for a few minutes before wiping.
4. Bathroom and laundry tiles, including mould.
5. Shower alcoves, grout and glass screens.
6. Bathtubs - fabulous to remove grime and tub rings as well as leave a streak free shine.
7. Mirrors – vinegar dries quickly and won't leave streaks.
8. Drain Cleaner
9. In and Around the toilet, (fantastic for urine smells).
10. Bathroom and laundry sinks with soap scum and build up – Sprinkle a little bicarb soda or salt if you need a bit of abrasive action.
11. Stainless steel surfaces and appliances to remove finger prints and grease.
12. Tiled floors and linoleum – Add ¼ to 1/3 of a cup of undiluted lemon vinegar.
13. Sweat stains - Spray recipe 1 on marked areas for a natural laundry stain treater for sweat marks, especially on whites.
14. Eliminate Jar Smells – add the mixture and dilute with water, allow it to soak and don't forget to do the lids too.
15. Removing sticky labels.
16. Shine up your vases and glassware by wiping or rinsing in the vinegar solution.
17. Vases with scum can be left to soak with the mixture and when you are ready to scrub, add bicarb or salt to remove any hard to clean areas.

Careful -

Avoid frequently using on rubber, wood and delicate surfaces and if unsure do a test patch first.

Remember – only use lemon on grout as it may discolour over time.

Vinegar isn't very friendly in the garden so don't pour your cleaning waste or water on anything except weeds.

Also get maximum use out of your cloths by washing them in the washing machine and reusing them, avoiding using harsh chemicals that would degrade them.

Clothes Washing

Wash your clothes only when needed, at 30°, using a full load on a short cycle. Then air dry if possible. This will of course reduce your consumption of electricity, water and soap, but it will also make your clothes and your washing machine last longer. Plus it'll help to reduce the amount of microfibres in the ocean.

Soap Nuts

These deserve a headline of their own! We switched to soap nuts for a non toxic alternative to horrible washing chemicals and we noticed they made such a difference to our skin. My sister, who isn't zero waste, used them when she came to visit and she commented that her clothes were smelling much fresher, like nature. These amazing little berries also last for up to 4 washes and you only need 5-7 berries in your wash, so they save a lot of money too!

Switch to Energy Efficient Light Bulbs

Switching to energy efficient LED light bulbs throughout your home is a great way to be more energy efficient. They can cost a little more than standard light bulbs but they last a lot longer and will save you money in the long run as well as saving the planet.

If everyone switched to energy saving light bulbs the world would save around £90 billion per year on energy costs! Wow!

Junk Mail

Put a sign on your door or letterbox

Put a 'no junk mail' sign on your door to help stop junk mail.

You can make a sign yourself - write "No commercial leaflets". You can also write "No free newspapers" or "Yes free newspapers" depending on whether you still want to get free newspapers.

It isn't a good idea to rely on signs alone as some delivery people might ignore them.

Also, be warned, I actually got abuse from a few postpersons for doing this! I guess it triggers those that don't care as much for the planet.

Contact Royal Mail

You can tell Royal Mail to stop delivering leaflets and brochures to your address.

You need to download this form from the royal mail website http://www.royalmail.com/sites/default/files/D2D-Opt-Out-Application-Form-2015.pdf

Fill it in and send it to the address on the form.

Royal Mail will send you a copy of the form if you can't print it yourself. You can phone or email them:

Royal Mail Door to Door opt out

Phone: 03457 740 740

Calls usually cost up to 55p a minute from mobiles and up to 9p a minute from landlines. It should be free if you have a contract that includes calls to landlines - check with your supplier if you're not sure.

Email: optout@royalmail.com

You'll stop getting unaddressed junk mail within 6 weeks of Royal Mail receiving your form. They'll stop delivering unaddressed mail to you for 2 years - then you'll need to fill in another form.

Register with the 'Your Choice' scheme

Registering with the Direct Marketing Association's 'Your Choice' scheme will help reduce the amount of marketing junk mail you get.

Contact DMA and ask them to send you an opt out form.

Direct Marketing Association

DMA House

70 Margaret Street

London

W1W 8SS

Telephone: 020 7291 3300

Calls usually cost up to 55p a minute from mobiles and up to 13p a minute from landlines. It should be free if you have a contract that includes calls to landlines - check with your supplier if you're not sure.

yourchoice@dma.org.uk

You'll start getting less junk mail in about 12 weeks. After 2 years you'll need to fill in another form.

Register with the Mailing Preference Service

Register with the Mailing Preference Service (MPS) - this will stop advertising material that's addressed to you personally.

You can register online at the MPS website, or you can contact them by phone or email.

Mailing Preference Service

020 7291 3310

Calls usually cost up to 55p a minute from mobiles and up to 13p a minute from landlines. It should be free if you have a contract that includes calls to landlines - check with your supplier if you're not sure.

You can't register if you have a PO box or business address, or if you live in Ireland.

You should start to notice a difference soon after registering - it can take up to 4 months for the service to be fully effective. Your details will stay on the service once you've registered - you should let MPS know if your details change or you move.

Stop charity marketing mail

You can contact the Fundraising Preference Service if you want to stop getting marketing mail from a charity registered in England, Wales or Northern Ireland. They'll tell them to remove your contact details within 28 days. The best way to do this is to register on their website. You can register over the phone if you prefer.

Fundraising Preference Service

Telephone: 0300 3033 517

Monday to Friday, 8.30am to 5.30pm

Saturday, 9am to midday

Calls usually cost up to 40p a minute from mobiles and up to 10p a minute from landlines. It should be free if you have a contract that includes calls to landlines - check with your supplier if you're not sure.

You'll need to have your contact details and the charity name or registration number to hand.

You can contact the Scottish Fundraising Standards Panel if you want to stop getting marketing mail from a charity registered in Scotland. You can get more information on their website.

Contact your electoral registration office

You can search for your local electoral registration office on GOV.UK.

Ask them to take your details off the 'open register' - this is a list of people and addresses that can be bought and used for sending junk mail.

You can choose for your details not to be added to the edited electoral register when you fill out an electoral registration form. Tick the box that says "op out" of the open register.

Contact the sender directly

If you want to stop getting mail from a particular sender, contact them directly. Include:

- your full name and address
- the date
- this sentence: "Please stop processing my personal data for direct marketing purposes in accordance with Article 21 of the General Data Protection Regulations."
- a reasonable date that you want the organisation to stop sending you mail - Article 21 says they should do this within 1 month

Return to sender

If you get junk mail with a return address on the envelope, you should:

1. Write "unsolicited mail, return to sender" on the envelope.
2. Post it - you don't have to pay.

This won't guarantee that you won't get any more junk mail, but it's a way of letting the company know that you don't want any more mail.

Children

We can teach our children to be eco friendly mostly by role modelling, but also give them age-appropriate reasons why we need to take care of our planet and our health.

Spend time outside in nature

I believe the best way to start your child on the eco-conscious journey is to get them loving the outdoors. Instead of keeping them cooped up in the house when there is fair weather, take them to places like parks to play, for bike rides and on hiking trails or even for camping, let them climb trees, let them play in the mud and embrace nature. In the process, your children will associate the outdoors with fun, and then it will be easy for them to have respect for and adopt habits that conserve the environment.

Recycling

If there is one activity that helps keep our surroundings clean it is recycling because it cuts down significantly on the amount of waste we let loose on the environment especially non-biodegradable waste. Make recycling a habit in your home and teach your children how to do it. Teach them to separate rubbish and have different bins for each type.

Conserve water and food

Teaching children how to preserve food and beverages in the fridge or freezer is a great way to avoid throwing away huge amounts of food leftovers. It's a waste of resources to find out that your kids left fruit on a counter to spoil.

Teaching our kids to turn off the tap when they are done with one task is pretty simple. Let's not waste water between brushing your teeth, washing hands or shampooing whilst showering.

Encourage children to use environmentally friendly means of transport

We don't have to use the car all the time when going from place to place. Teach the children to walk or ride a bike to a place they need to go to that isn't far away from home. It's a great way to save fuel and also minimise air pollution. It's good to explain the effects of air pollution to our children and why taking a bike, running or walking to a place nearby is a good choice for the environment.

Love of animals

Animals are affected by pollution in a big way so it's good to explain this to our children. The best way to raise an eco-conscious kid that loves animals is to take them to places with animals from a young age. You could also start by having a pet when they are children, even a little pet like a hamster encourages them to take care of animals and be responsible with the environment.

Involve the children in activities with the family pet such as taking them for walks and feeding them. You can also take them to animal rescue centres where they will see lots of other animals. Then explain to them how important it is to keep the environment clean as waste harms animals and makes it unsafe for them to live in their habitats. Give examples such as waste in water that harms fish. When you instil this knowledge in your children at an early age, it will help them to grow up to be responsible adults, being aware of how they treat their surroundings.

Gardening

A great way to raise eco conscious children is to teach them sustainable gardening. Growing things and playing in soil is a great way to get them appreciating nature. During your activities teach them how to create manure using recycled waste. Show them how that produces food for the plants. Get them to take biodegradable waste to the compost heap themselves. They will love it because they will see the benefits in the growth of the vegetables they plant.

Tree planting

It's important to instill the love of trees in your children from an early age. Trees are super important for the environment, and your eco-conscious child needs to know this fact. A fun way to do this is to plant a tree for every one of their birthdays. Also take them along for tree planting exercises set up by different organizations especially during days such as Earth Day. In the process teach them how trees work well for the environment and why it is important not to cut them down. You can also have an art project where you cut out pictures of different types of trees found in your area and explain to them a few characteristics about them. In this way, you will also build their knowledge of their environment.

DIY projects

Do DIY projects that involve recycling or reusing things such as cardboard boxes, toilet rolls, plastic bottle and even glass ones. It's a fun way to keep them busy and also show them that trash can become art or turn into something useful. There are lots of projects online that you can use to teach your children this lesson. The idea is to show them that not everything goes in the bin and to find ways to reuse something before throwing it away.

Cleaning

Teach your children to pick up after themselves and dispose of waste responsibly. Have you seen adults that throw rubbish around in public? They probably weren't taught the importance of proper waste disposal in their childhood. We want our children to always know that littering is wrong. They will model us, but it's also good to teach them how to clear up after themselves and which item goes in which bin.

Library

Instead of buying books, magazines and DVD's, visit the library and borrow these resources. Libraries are a valuable resource for sustainable living and an exciting place for children to explore. My children still get super excited about a trip to the library.

Children can also get involved in the activities hosted in the library.

It's also vital that we use our library facilities so that they don't get shut down. We had to protest and petition to save our local library from being shut down, and it was closed for months, but the council finally opened it again, part-time.

Community efforts and projects

Involve other parents in activities that help to save the environment. You could organise to share cars to minimise the use of cars, too. Arrange tree planting or gardening exercises where you bring your children together and teach them about the importance of plant life. Play dates where you make objects or use toys made from recycled materials such as cardboard boxes and bottles are also a lot of fun.

Involving the children in activities such as upcycling furniture will show them the importance of re-using items. Their brains will take in information faster during fun activities.

Non Toxic Biodegradable Slime Making

Avoid buying the smelly toxic slime from the shop in all its plastic packaging and get the kids busy making their own. You don't need glue or borax either!

Ingredients

1/2 cup Cornflour
18-20 Teaspoons of Water
Food Colour or Natural Colour

Instructions

1. Measure out 1/2 a cup of cornflour
2. Place some water in a small dish and add some colour

3. This makes it much easier to stir and less chance of staining your hands or clothes in the process.
4. I don't measure the water at this point, there's no need to.
5. Place cornflour in a dish and add 10-12 teaspoons of your coloured water on top.
6. Mix together with a butter knife or spoon. It should start to look a little crumbly.
7. Slowly add extra teaspoons of coloured water to the mix until you reach a gooey slimey consistency.

It's kind of like making cupcake icing really and only takes a minute or two to do.

If you made it too watery, just add a little more flour.

If it becomes too dry, especially after the kids have played with it for a while, just add a little more water.

How do you know if the slime is the right consistency?

When you scoop it up with a knife it should slowly drip down like a goopy mess. Then when you grab a bunch of it in your hand you should be able to roll a ball.

When you're rolling your hands together, it will keep its shape and as soon as you stop, it turns straight back to gooey slime and drips everywhere!

Make Biodegradable Glitter

This won't be exactly like "normal" glitter but it's a great option when you're avoiding micro-plastics!

It's very simple to make and can be glued onto pictures, craft projects and homemade cards. It's super easy to clean up afterwards and can be put in their bath at the end of the day.

Ingredients

1 Tablespoon of Epsom Salts

1 Tablespoon of Table Salt

4-5 drops of Liquid food Colour

When selecting ingredients, opt for those in the most recyclable packaging and as plastic-free as possible. Food colours usually come in plastic bottles, but these do last a long time and can be reused for so many things.

Instructions

1. Add the epsom salts to a glass container
2. Drip food colours in carefully and give it a little stir with a spoon
3. Pop on the lid and give it a really good shake
4. That's it :) You can use immediately but I find it best to allow it to air dry for around 30 minutes to an hour.

Biodegradable Non Toxic Play Dough

Ingredients

1 Cup of Plain Flour
1/2 Cup of Salt
2 tbsps. of Cream of Tartar
1 tbsp. of Cooking oil
1 Cup of Boiling Water
Food colouring

(You can halve or double the ingredients depending on the quantity you want to make)

Instructions

1. Mix all ingredients in a bowl or jug apart from the hot water and colour. The kids can help measure and mix, but the next steps are best done by grownups

2. Pour the hot water into a jug to add the colour. This helps to get an even colour throughout. Keep in mind that the colour of the water will be much stronger than the colour of the play dough. You can always add more colour during mixing if you feel it's not quite right.

3. Add the water and stir. I usually use a butter knife to stir as it's easier to wipe off than spoons.

4. Once it is all starting to stick together and looks like dough, pop it onto the kitchen counter and knead briefly. Be careful, as the playdough will still be hot in the centre.

If you find it is still too sticky, add a little more flour and knead through.

5. Wait about ten minutes until it cools enough for little hands to play with.

To ensure it lasts, and to prevent it drying out, keep it in a closed container when not in use. If it does happen to get left out overnight, knead it and place into the storage container and you can add a few drops of cooking oil to it if it's too dry.

If you're making this for several kids, double the recipe so there is plenty to go around.

Cloud Dough (kind of like kinetic sand)

Ingredients

4 Cups of Unbleached Flour

3/4 of a Cup of liquid Coconut Oil

1 teaspoon of Vanilla Essence

Liquid Food Colouring

Instructions

Add the flour to a large mixing bowl

Mix in the oil and vanilla essence and stir through to create your basic Cloud Dough

Choose your colours. We use rainbow colours. Add around 1/2 a teaspoon of each colour at a time and use a fork to mix the colour around.

Due to the consistency of cloud dough, the food colour won't blend evenly and will give you pretty speckles of colour.

You can use any cooking oil and flour, so if you have stale items, this is such a fun way to use them up rather than throwing them away.

I chose a local flour that I can buy in bulk recyclable paper bags and check your local bulk store for package free oil to make this zero waste.

As there is no heat involved, kids can help with the entire process.

Use items from your kitchen to create fun shapes or just mould with your hands. Store in tubs when not in use.

Non Toxic Fizzy Bath Bombs

For moulds you can use silicone, plastic or metal cupcake trays, but also cookie cutters and ice cube trays make fun shapes, or just shape into balls by hand. I used to think I needed to buy special bath bomb moulds to make these and that was a huge put off for me as many were made of flimsy looking plastic and I couldn't justify purchasing the metal ones.

This recipe takes around 10 minutes to make and you'll get 5-6 extra large bath bombs or fills two ice cube trays.

Ingredients

1 cup of bicarbonate soda
1/2 cup of Citric acid
1/2 cup of epsom salts
4 teaspoons of liquid coconut oil
1 teaspoon of water
A dash of food colour (approx. 1ml or 20 drops)
Essential oils (Optional)

Tools

A whisk or spoon
Mixing bowl
Measuring cup
Teaspoon
Moulds - muffin trays, ice cube trays, chocolate moulds, cookie cutters (or you can shape by hand) Silicone moulds are best as they are simple to clean and easy to remove.
Clean the surface with a bit of elbow room

Instructions

Mix all dry ingredients thoroughly in a mixing bowl
In a separate jug (or cup), mix liquids (oil, water, food colour and essential oil if using)
For fizzy fish bombs, I use about 10 drops each of blue and green liquid food colour and 7 drops of essential oils

1. Add this liquid really slowly to your dry ingredients.
2. If it begins to fizz, slow down as this could compromise the fizz level of your bath. Whisk or stir quickly with a spoon until it looks crumbly but is soft and a bit squishy.
3. Place into moulds or cookie cutters (or shape by hand), and ensure you press firmly. They need to be well compressed. They may expand a little while drying.
4. Allow to dry for at least 2-4 hours before removing from the moulds. 4 hours after making they are ready to use :)

These really fizz up nicely and if you have any small kids toys, these can be hidden inside the bath bombs when shaping. Kids get super excited waiting to see what is hidden inside (even if they already know what it is).

Suggested Fragrances

There are so many essential oils that you could use alone, or in combination. Here are some suggestions:

Lavender
Bergamot & lime
Sandalwood & Ylang Ylang
Peppermint
Rosemary & Lemon balm
Geranium & Rose
Jasmine

Reusable Nappies

Single-use nappies end up in landfill and take a very long time to decompose, they also create a lot of sulphur which is toxic for our environment.

Switching to cloth diapers can be a great way to reduce your family's daily waste, especially if you have a few children, they can be passed down.

These days washable nappies are available in bright colours and patterns and with easy to remove liners. They also have adjustable snap fasteners so you can use them as your child grows.

Reuse and recharge

Buy rechargeable batteries for your kids' electronics and toys and teach them how to care for and recharge them. This reduces garbage and keeps toxic metals, like mercury, out of landfills.

Pass it on

Ask kids to gather their toys, books, clothes, and other goods that they no longer use or want for donation to local charities. Take them along for the drop-off so they can see how groups like the Salvation Army, or charity shops use donations to help others.

Watching cartoons and films on saving the environment

Use your child's TV time to teach them something about the environment. There are lots of videos or cartoons for kids that show the importance of preserving the environment. You could get comic books and storybooks that teach them how to preserve the environment as well. This will help them understand what it means to respect the environment and how to do it.

Here's a list of cartoons and films to inspire your young eco warrior -

Sesame Street https://www.youtube.com/watch?v=kTWE18VPj-c#

Sesame Street is one of the top kids' education shows in the world. It's been around for over 40 years and Elmo, Cookie Monster, Big Bir, Bert and Ernie have always taught kids to care for their environment - whether that's by saving water or recycling. They usually do it with the help of a celebrity.

Captain Planet https://captainplanetfoundation.org/about/our-story/#

"Earth! Wind! Fire! Water! Heart! Go Planet! By your powers combined I am Captain Planet!" Many kids of the 90s will remember the opening lines of the theme song to this short-lived animated environmental show.

"Captain Planet and the Planeteers" saw Gaia - the spirit of earth - awakened from a long slumber by eco villain Hoggish Greedly. To her horror, Gaia realises the planet is badly damaged and sends out five magic rings across the continents. Four of the rings control an element of nature and one controls the element "heart."

The five eco warriors who receive the rings are called the Planeteers and their job is to fight environmental destruction and raise awareness about it. The Planeteers can combine their powers to summon green superhero Captain Planet, whose catchphrase is "the power is yours," reminding viewers that they have the power to save the planet. Each episode ends with a discussion of an environmental problem, telling viewers how they could be part of the solution rather than a contributor.

The Octonauts

In the modern British animated series, "The Octonauts" children learn about the ocean, sea animals and how to protect the environment.

In "The Octonauts," Barnacles, the brave polar bear, leads his seven-person team of anthropomorphized animals in exploring the ocean depths. While the technology in the show is sci-fi, the team encounters unusual but real sea creatures in each episode. They often have to discover a biological or behavioural fact about the creature in order to get it or themselves out of danger.

Marine biologists Lara A. Ferry-Graham and Michael H. Graham helped out on the series to ensure it was accurate while being fun for kids. The scientists also advised on children's film classic "Finding Nemo."

Bill Nye the science guy

In another 90s U.S. series that was also popular around the world, wacky scientist Bill Nye made science easy and fun for preteens. Nye explained a range of subjects, including ecology and environmental science with the aid of crazy experiments.

Though the TV show was cancelled in 1998, the science guy still appears in science videos, explaining subjects like climate change.

Bambi

Bambi," first released by Walt Disney in 1942, is a classic film that inspired a generation of conservationists. (Former Beatle Paul McCartney has credited the shooting death of Bambi's mother for his initial interest in animal rights, according to BBC.) **Warning:** the death of Bambi's mother is traumatic!

Avatar

The director James Cameron spent $237 million and almost a decade making an incomparable blockbuster plea for environmental protection and respect for indigenous lands. That's a truly remarkable effort. It is the most expensive environmental advocacy effort in human history—and wildly successful.

That Avatar earned almost $3 billion worldwide—making it the most successful film in history—only reinforces the importance of the film. Fifty years from now, Avatar may be remembered as having played an important cultural role in the era of climate change politics.

The highest-grossing movie of all time also happens to be an environmental call to arms by a director who has compared climate change to "the threat the United States faced in World War II."

The Wombles

For people from the UK, "The Wombles" is probably one of the most enduring childhood memories. These pointy-nosed, furry fictional animals first appeared in a series of children's novels in the late 1960s and in the 1970s in a BBC stop-motion animation show. The thrifty Wombles lived in a burrow on Wimbledon Common in London, England where they helped the environment by collecting and recycling rubbish. Their motto was "Make good use of bad rubbish" and their love of recycling was way ahead of its time.

Parties

A few Mums I spoke to recently wanted to do an eco friendly birthday party but just had no clue where to start because it's kind of ingrained in us to send party attendees off with plastic party bags filled with plastic toys, and use plastic cutlery.

Food Ideas for Birthday Party Bags

Popcorn
Mini Sultana boxes
Homemade treats like cookies
Boxed Smarties
Foil wrapped chocolate
Bulk stores are your friend
Lollies
Rice crackers and seaweed snacks
Soya crackers
Veggie chips
Dried fruits
Yoghurt balls
Dried apricot treats

Eco Friendly Alternatives to the Plastic Party Bag

Paper lunch bags

Small up-cycled cardboard boxes

DIY newspaper bags

Fabric bags

Jars

Paper cones

Ice-cream cones

Reusable pouches

Work

Don't worry about any naysayers at work, or people that seem to not care. Be the shining light and just be a good example, they'll follow eventually. Here are a few suggestions for doing zero waste in the workplace:

Take a packed lunch

In my Mums workplace no one cares about Zero Waste or being conscious about their decisions, so she found it really hard when she first started reducing her waste and eating healthier. She wanted to tell everyone about this better way of life, but they weren't interested, in fact some of them were even horrible, argued, or dismissed her as the crazy lady. But she started taking delicious home-made soups and home-made plant based curries in reusable containers, the aroma filled the staff room and people started asking what she was eating, she'd let them try a bit and they were shocked at how delicious good food could be, much better than their shop bought plastic packed sandwiches. She started sharing her snacks as well, homemade roasted chickpeas and nuts mostly, in glass containers. This got her colleagues interested because before they just assumed that zero waste and toxin free living was a far removed idealism that none of them could manage, only the crazy lady. Well, from just sharing her packed lunch in her bento box, 5 of her colleagues now have their own bento box, reusable water bottle, use hand soap and don't wrap anything in plastic cling film anymore! Result!

Encourage the use of reusable fountain pens

When it comes to stationary you have the option of either using a refillable fountain pen or sticking with pencils because they are biodegradable. Fountain pens are totally back in fashion, so let's make sure that trend stays. You can purchase ink in glass containers, which is another zero waste win! As for notebooks and other office essentials, try to get a hold of things second hand first, or only purchase items if you know they will be reused again and again. Or as a last resort, purchase recycled items, like notebooks.

Go paper free

With the rise of cloud storage services the need for paper and printers in the workplace has rapidly decreased in past two decades.

Diminish the temptation to print by reducing the amount of printers in your office and donating them to a local school or non-profit organization in need.

Unplug electricals

Don't just power down your devices before leaving — power is still being consumed if it's plugged in.

Install motion-activated light switches

Improve energy efficiency immensely by installing or swapping in motion-activated light switches. These are a particularly great solution for conference rooms, since they are often separated from the central office space and are less frequently used.

Rather than relying on employees to switch-off the lights as they leave or enter, motion-activated lights are human proof: they save energy automatically.

Office Furniture

Thankfully there are options to buy furniture made from more sustainable materials these days, such as bamboo and cork. Or, instead of buying new you could go for an amazing upcycled desk.

Keep the blinds open during daylight

If your work space is equipped with windows, this is really a no-brainer. 20% of the electricity consumed in the UK is used for lighting in businesses. Cut your environment and financial losses by opening the blinds and letting the natural daylight pour into your workplace. Switch to LED light bulbs too.

Use bamboo cloths rather than paper towels

In most workplaces, paper towels are the go-to method for cleaning up spills of any kind. There are, however, many inexpensive eco-friendly alternatives.

Get a stack of bamboo cleaning cloths and attack that spill with some eco enthusiasm!

Buy in bulk

Some people think single-serving packets of sugar, cream, salt or pepper are fancier and more convenient than buying an entire container.

In reality, buying these items in bulk saves money and unnecessary waste, they're also generally easier to find in the shop. It also will prevent employees from wasting the product itself, rather than using one and half creams packets, employees can pour out the exact amount they desire from the carton of cream.

That said, we don't recommend you keep a messy paper bag of sugar on your kitchen countertop. Instead, find a reusable glass jar, fill it with sugar and small spoon, and put it next to the office coffee pot.

Clearly label rubbish, recycling and compost bins

Whilst most people know recycling is a good thing, many people are still confused as to what can and can't, go into the recycling bin.

Help your employees/colleagues become eco-warriors by clearly labeling which type of waste goes where. It's also helpful to send out a quick refresher email or video every now and then so that employees stay mindful.

If your facility does not offer composting as part of its waste management service, do some research to find a third party composting service in your area.

Initiate a BYORB policy (Bring Your Own Reusable Bottle)

Bringing your own bottle, tumbler, or coffee mug to work is one of the easiest and healthiest eco-initiatives to have in the office.

Skin Care

My go-to product for skin care is actually just our organic black soap because it's anti-bacterial, anti-fungal, anti-inflammatory and very moisturising. This was a lifesaver for my family when we suffered from eczema, so we still use it as a daily soap to keep the eczema at bay but also because it produces beautiful skin. If you've got dry, itchy or flaky skin I can't recommend our organic black soap enough. So many of our customers have used it to help with their eczema after trying everything else, even after using steroid creams to no avail.

Make-up

I find Arbonne makeup to be the best green makeup, it's toxin free and has the same professional finish I was used to as a professional model, I actually didn't wear makeup until I found something I really loved that was toxin free, I still don't wear it much because I believe in skin care over covering up, but it's nice to wear it when you feel like being a bit creative, or even for special occasions. The liquid foundation is lovely and I like to use the lip liners as an all over lipstick because they don't budge all day and they're super moisturising. You can get them here - http://KimCalera.arbonne.com/

I use Terracycle's Zero Waste Beauty Box to recycle. When you receive your box fill it with the appropriate waste items. Once full, take your box to any UPS location or schedule a collection to send your items back to TerraCycle using the pre-paid UPS shipping label which is already affixed to your box. When they receive your box, they safely recycle all of the collected materials. Here's a link to their service. https://www.terracycle.com/en-GB/zero_waste_boxes/personal-care-and-beauty-waste

Some of TerraCycle's services are free, and some cost because many types of plastic involve costs to break down and reuse, so why not ask your place of work or study to invest in a box and everyone gets to fill it up?

Lush also have foundation bars, concealer bars and more zero waste cosmetics available.

Safety Razor

If we think about zero waste and the "olden days" it's very similar.

Back then, people didn't have the materials we have today. So, instead they reused everything and found lots of different uses for the same item. This would save them money, but also they didn't have plastic convenience in those days, maybe that's why they call it the good ol' days...

So for hair removal they didn't have waxing and expensive laser hair removal, a safety razor was all the rage.

I'll be honest with you, when I first got my safety razor I was scared to use it! When I finally plucked up the courage to try it, I was shaking because I was so nervous. Looking back now, I find that hilarious because it really isn't that bad at all. There was no blood and it gave a really smooth result.

Reusable Cloth Makeup Removal Pads

We make bamboo reusable makeup pads and sell them at Honest Miracle. You use them just like "normal" makeup removal pads and then you just pop them in your washing machine. I actually think they're more comfortable too. I love bamboo.

DIY

You could even try your hand at making some DIY products. Coconut oil is a great body moisturiser. Try some apple cider vinegar for blemishes.

Zero Waste On The Go

Takeaway Coffee

IF you have the option to drink it in a café, simply request a real cup. Most places will provide real cups if asked, although some will try to refuse you as there seems to be a thing about avoiding dirty dishes these days.

If you're a regular on the go coffee drinker, and don't have a reusable cup yet, this is one simple action you can take right now to make a huge difference in the world.

You don't need to buy a fancy one and I've heard of people using jam jars with a cosy to protect from the heat and yes, you can just bring an ordinary coffee mug without a lid.

Did you know some cafes offer a discount on your coffee if you bring your own cup? Leon, wholefoods, Cafe Nero, Costa, Starbucks and M&S Cafe are some that offer a discount.

If you do find yourself caught out, ask for no lid and be sure to recycle your cup re-sponsibly.

Avoid Bottled Water

Purchasing bottled water and drinks is pretty easy to avoid, simply bring your own or keep your eye out for public water fountains. Lately I've spotted more water fountains that have been designed to refill water bottles too, which is brilliant. I am yet to find a café or pub that doesn't have fresh un-bottled water available for free that you can either drink or you can refill your own bottle.

My daughter was totally shocked when I told her that when I was a child we didn't take water everywhere, we just used drinking fountains and occasionally had a flask in our lunch box. We had a drink before we went somewhere, or after we got home and we never came close to dying of dehydration. These days, it's become the norm to carry water or buy water.

I bet you have a few reusable bottles sitting in a drawer or cupboard in your kitchen and you can repurpose jars and glass bottles you already have.

There are so many types of reusable drink bottles available, find one that's the right size, shape and material for you.

Reusable fizzy drinks

Have you got a soda stream? This is ideal for making your own bubbly drinks and taking them with you. I don't know just how environmentally friendly they are overall, but if it replaces constant disposables it seems like a fabulous step in the right direction. On their website it says that one can of bubbles replaces 40 bottles or 180 aluminium cans, just imagine the impact of less drink miles, packaging. We don't really drink fizzy drinks, so I have no clue how they work and how good they are, but lots of people have them and aren't using them, so you can probably get them cheaply second hand too. If you don't fancy making your own then you could just buy fizzy drinks in glass bottles.

Reusable straws

Plastic straws seem to be the most mainstream zero waste swap right now, everyone's talking about how bad they are. They are certainly damaging to the environment and sea life, but they probably aren't the worst thing happening to the environment. The big issue is that they are used for such a short time and last forever, and often we don't even want one but we're served them.
The simplest option is just to say no and request no straws.

I get that some people really need straws and this includes people with a disability, difficulty drinking or like my friend, has super sensitive teeth. Choose reusable straws.

There are a lot of different reusable straws you can get, ranging from bamboo, glass, stainless steel and silicone. You can find all different widths, lengths, straight ones or bent ones, plus it's easy to find straw cleaners. (If you're only using it for water, you might not even need a straw cleaner). We have a reusable stainless steel straws + cleaner set at Honest Miracle.

Travel

Holidaying

I know some zero wasters that only holiday in the UK and that's fine, it's a personal choice, but we are worldschoolers and we do go abroad, but we have started going by train! It's much better than taking a plane because there's less fuss and you can take pets, plus depending on where you go, it can be faster than taking the plane. But of course, the huge benefit is that it lessens the environmental impact. Did you know that travelling from London to Paris by train produces 91% less CO2, according to Eurostar and Seat61.

If you do choose to go by plane (or have to, let's be realistic, sometimes things happen and we need to adjust, things like family emergencies etc) you can do a few things to reduce your impact on the environment like -

Carrying snacks or a meal

Most airlines will allow you to carry food as long as they do not contain ingredients which might cause someone to have an anaphylactic shock.

I've taken fruit on planes and I've even seen people take sandwiches which you could keep in a lunchbox in your bag. For some reason that didn't occur to me until I seen it, I thought that because of tough security regulations we wouldn't be able to take "actual food" on an aeroplane, but you can :)

Reusable Biodegradable Cutlery

I also recommend bamboo reusable cutlery when travelling. In Europe I am always shocked at how much plastic cutlery gets handed out and usually on planes and trains they'll serve you plastic cutlery with meals (in economy anyway, I wouldn't know about any other carriages). We actually sell bamboo reusable cutlery in travel pouches at HonestMiracle.com

Everyday Travel

Ditch the Car

Obviously, this one won't work for everyone but if you live in a large town or city that has a good public transport system then why not consider ditching your car completely.

Cars emissions have a huge impact on our environment so switching to public transport definitely makes you a sustainable living warrior.

It's pretty easy to bike around if your area has cycling lanes, too. Sometimes we just get into a habit and are scared of change but whenever I bike around I'm always shocked at how fun and relaxing it is, and can't wait to do it again!

When it's cold outside

Buy Thicker Curtains

Having thicker curtains helps to keep the heat trapped in your home. If you need to buy new curtains then perhaps consider thicker ones. Alternatively, if you already have curtains in your home you can buy material that you can attach to the back to make them thicker.

I'd also highly recommend opening up your curtains during the day to get the most solar heat from the sun. Then close them again at dusk to trap in the heat you've collected during the day.

Seal Any Gaps

Did you know that 40% of your homes heat is lost through your windows, doors and floors? A couple of things to consider are -

- Self-adhesive strip draft sealers
- Under door sweep stripping
- Heavy-duty rubber seals
- Storm guards

You can buy these inexpensively at DIY stores or online.

Insulate

Insulating your home can have a huge impact on your carbon footprint. Houses with better insulation use less heating which means lower energy bills and less energy being used.

Insulate your water tank

If you have a water tank at home it's a good idea to insulate it to help stop heat loss and keep your water hotter for longer.

Install double Glazing

Similar to insulating your home it's also a good idea to have double glazing in your windows if possible. If double glazing isn't an option for you then look into installing draft excluders around your window seals. There are lots of different types available at various different price points.

Close your damper in your fireplace (ONLY IF YOU'RE NOT USING IT)

Finding and sealing all the gaps in your house should be enough. However, if you still feel a cold draft, your fireplace could be the culprit. It might be the damper in your fireplace.

Having an open damper is similar to having an open window in your home. When you're using the fireplace, the open damper prevents smoke from filling up your house. But when it's not in use, it becomes an entryway for cold air to come in.

It's actually something that a lot of people don't notice, but if you want your home to be truly energy-efficient and you don't want to waste even a tiny fraction of heat, better check the damper too.

Please remember to open it before you use your fireplace though because the smoke can kill.

When it's hot outside

Instead of reaching for the fans, or air conditioning here are a few things we can do to keep our homes cool in summer:

Thermal Curtains

When it's hot outside thermal curtains are your friend! Keep them closed and keep the heat from the sun out.

Keep windows closed during the day

I see so many people with their windows wide open in heatwaves, and I used to do it myself, but after researching and trial and error, I discovered that it's best to keep our windows closed during the day and then open them in the evening when it gets cooler.

Keep doors closed

Try and stay in one room if you can and make that your "cool room", so close all the doors in the house and roll up towels to place at the bottom of the door to stop heat from getting in.

Put bowls of cold water in the room you're in

The cold water helps to cool the air and it also adds in some moisture which helps to cool it more

Water spray

Spray water mist over your body frequently to keep you cool

Energy efficient light bulbs

These will reduce the heat build up in your house and they'll also save you money and save the environment.

Solar Panels

If you plan to live in your home for a long time then it's definitely worth investing in some solar panels for your roof.

Try to find a company like EDF that use Feed-In tariffs. *'A payment made to households or businesses generating their own electricity through the use of methods that do not contribute to the depletion of natural resources, proportional to the amount of power generated.'*

Here are some of the benefits of a Feed-In tariff...

- Your energy provider will pay you a set rate for each unit of energy you generate.
- You'll then get another amount from your FiT Licensee for any extra energy that you export back to the grid.
- You will save money on your electricity bills.

Choose Green Energy

I have a few friends that live off grid, but for the rest of us, if you're not interested in getting solar panels, then the next best thing is using green energy. We are with Bulb who provide 100% green energy, our bills also reduced SIGNIFICANTLY when we switched, I'm talking a saving of over £500 a year. **You can get £50 from Bulb for free using my link** - www.bulb.me/kimberleys7244

You can use the free £50 towards your energy bill or ask them to send you it in cash.

Downsize

Downsizing is obviously quite a big lifestyle change but it can have a huge positive impact on your life.

Downsizing has not only reduced our costs but we also have a lot less impact on the planet. Plus we have less to maintain which means fewer resources being used.

It's amazing how freeing downsizing is too.

Gifting

Children

Books - Wild Tribe Heroes series of books on plastic pollution and palm oil - https://wildtribeheroes.com/

The Perfectly Wonky Carrot book about food waste and body image - https://new-many.co.uk/books/theperfectlywonkycarrot/

The Tantrum That Saved The World book about a little girl who channels tantrum power into saving the world from climate change.

Ask people what they want

The best way to avoid gifts going to waste is to give people something they really want. The element of surprise can be part of the fun, but it's also a major cause of waste, as it can be difficult to accurately guess someone else's tastes, and these are the gifts that end up stored up in bottom drawers for the rest of the year because they'd feel too bad to give them away. If you think your recipient would like it, why not ask them to give you a clue about what they want? It doesn't have to be super-specific – one year my Mum asked for a watch with a dark face and light hands, and I chose the rest.

Buy less

Do you need to buy for quite so many people? Do you need to buy so many things for each person? Some people receive more smellies than they are likely to use before next Christmas. One year we were given so many bath and shower smellies that wer-en't actually suitable for any of us (sensitive skin + toxins) and I felt so bad because I wanted to be honest so they didn't waste their money again, but also didn't want to come across as ungrateful.

Giving handmade

Handmade gifts are a popular choice for many zero wasters. They are definitely fun to make and lovely to receive, although they will only be better for the environment if the ingredients and methods used have less impact than their shop bought equivalents.

If you like to give handmade gifts, see if you can make use of anything that is currently going to waste, and if you're making food, choose methods which use the minimum amount of energy.

Gifting experiences

This is another popular choice among zero wasters, although again, obviously this only works if the experience is an environmentally friendly one. Maybe don't buy anyone a voucher for two at the local steakhouse. We usually request year passes to family attractions. Experiences don't have to cost money either, you can also offer your time or help with a project.

Gifts to enhance nature

The UK (along with much of the rest of the world) is facing something of an ecological crisis, with rapidly declining wildlife. If you know someone who is a nature lover, maybe they would enjoy something that attracts wildlife to the garden like butterfly and bug houses, bird feeders and wildflower. Some of these might also make good gifts for children.

Giving second-hand

Is this really such a no-no? There may be some people who would really mind receiving second-hand gifts, so if this sounds like any of your family/friends, maybe don't buy anything pre-loved for them. Otherwise, why not? Me and my Mum have even been swapping the same reusable wrapping bags back and forth for several years.

Recycled and upcycled gifts

Upcycled gifts are unique and reduce the demand for new stuff to be produced. They can be found at craft fairs. We have a lovely antique/upcycling shop near us, and upcycling is getting more popular now, so have a look to see if you have a local upcycling business too. Gifts made from recycled materials can be found at Oxfam (who also do organic and fair trade treats), The Eden Project, Protect the Planet as well. In fact, you can find many things available in recycled materials, such as clothes, socks, accessories etc. And don't forget recycled food – Toast do a selection of beers brewed from bread that would otherwise go to waste.

Honest Miracle's Customers Top Tips

We asked our eco community to give us their top tips for our readers and here's what some of them said:

Name: Jacqueline

Location: Northern Ireland

As a parent of young children I find that there is a lot of waste based around kids snacks. So many parents are buying drinks in single use plastic bottles or cartons, raisins in single use boxes, biscuits in single use wrappers etc.

I reduce this by-

1. Buying one reusable BPA free bottle and filling it with water and taking it everywhere we go (or could be filled with dilutable juice from one large bottle),

2. Buying a large bag of raisins from the baking aisle and putting a portion in a reusable tub to take with us on a day out/for lunch box etc and refilling each time from the large bag.

3. Buying large bags/packets of cheesy biscuits etc and taking a few out and into a reusable container for lunches, day trips etc.

This reduces single use waste significantly and also saves me so much money!!

Name: Charlotte

Location: Cambridgeshire

Plastic is polluting the oceans and it's now being found in fish, animals and even humans! It's very easy to reduce your plastic use. Here is what I started with:

1. Replacing plastic carrier bags with reusable bags.
2. Use stainless steel straws instead of plastic ones (I drink a lot of lemon water and it's good for your health but not good for your teeth, so straws are required)
3. Reusable sanitary pads to be used as incontinence pads for the leaks you get after having children. Before, I didn't realise how much plastic and waste is in disposables!

4. Don't buy plastic bottles, only reusable stainless steel or glass bottles.

Name: Judith

Location: Hampshire

I started my zero waste journey quite later in life after seeing a document on the television about the plastic in tea bags. My tips are:

1. Use loose leaf tea. There is so much plastic, bleach and toxins in normal tea bags. This is not good for the environment or for our health.
2. I got the reusable tea bags from Honest Miracle to make loose leaf tea drinking a bit more convenient. I know strainers can be a bit of a bother, especially to clean.
3. One of the easiest swaps is going back to bars of soap. There are so many lovely handmade soaps these days that it doesn't make sense for us to pollute the planet with all of this plastic. It's not even more convenient to use the plastic bottles of soap. I remember when they were brought out and I am quite ashamed to say that I fell for it but now I am back to my trusty old soap.

Easy Eco Shopping Guide

Here's a secret discount code for you and your loved ones to get 15% off anything in the shop - **ZWBOOKVIP** - just enter that at the checkout.

Swap cling film and foil for reusable toxin free food savers -

https://honestmiracle.com/products/reusable-toxin-free-food-savers

Swap toxic plastic and chemical ridden single use sanitary pads for reusable toxin free organic bamboo sanitary pads -

https://honestmiracle.com/products/reusable-sanitary-pads

Or a reusable menstrual cup -

https://honestmiracle.com/products/reusable-menstrual-cup-uk

Swap plastic toothbrushes for bamboo toothbrushes -

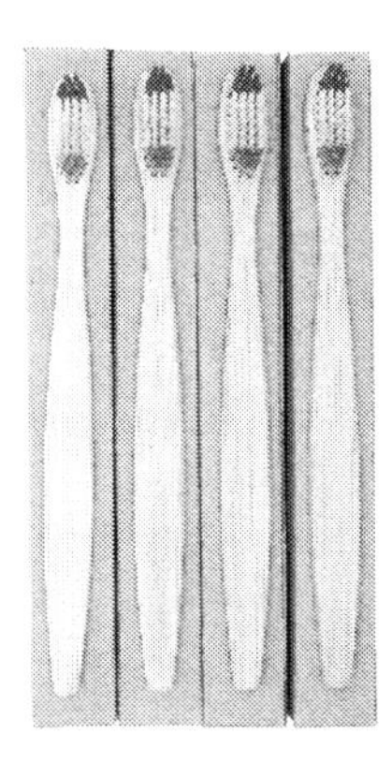

https://honestmiracle.com/collections/eco-friendly/products/eco-friendly-bamboo-tooth-brush-rainbow

Say no to plastic ridden tea bags and yes to Reusable Toxin Free Tea Bags -

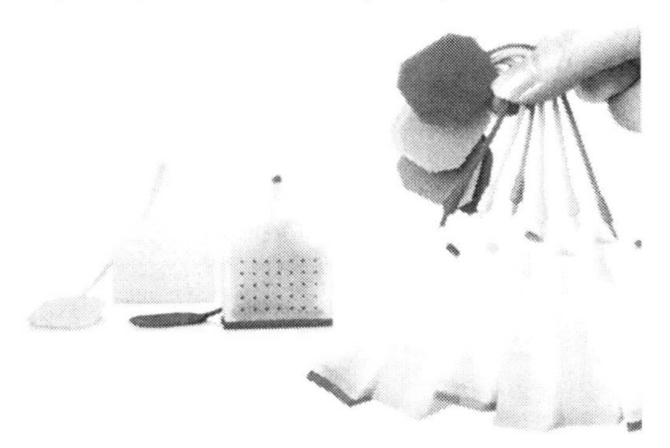

https://honestmiracle.com/products/reusable-toxin-free-tea-bags

Looking for a naturally antibacterial soap bar? One that soothes and moisturises yet deeply cleans? Check out Organic Raw Black Soap -

https://honestmiracle.com/products/organic-raw-black-soap

Refuse single use produce bags, use Reusable Toxin Free Produce Bags instead -

https://honestmiracle.com/products/eco-friendly-toxin-free-reusable-produce-bgs

Refuse plastic straws, get your own reusable stainless steel straws and brush cleaner set instead -

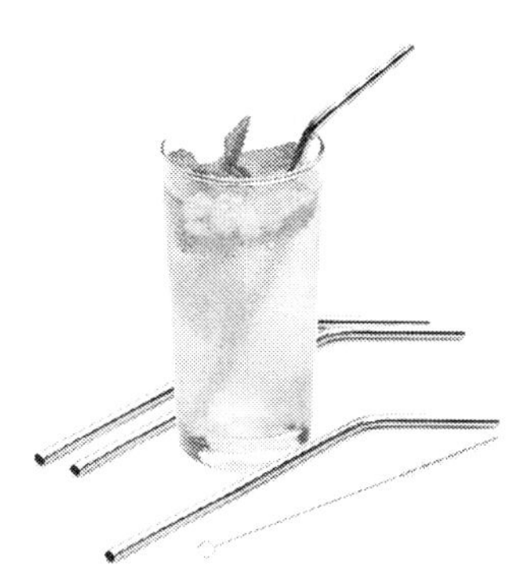

https://honestmiracle.com/products/eco-friendly-reusable-toxin-free-straws

Or a collapsible stainless steel straw -

https://honestmiracle.com/products/reusable-collapsible-travel-straw

Treat yourself to a reusable bamboo cutlery travel gift set to take around wherever you go -

https://honestmiracle.com/products/reusable-bamboo-travel-cutlery-gift-set

Here's a reusable shopping bag that you can keep on your keyring -

https://honestmiracle.com/products/eco-friendly-reusable-folding-shopping-keyring-bag

Swap face wipes for reusable organic bamboo make-up removal pads -

https://honestmiracle.com/collections/reusables/products/bamboo-breast-pads-nursing-pads

Show how proud you are to be an eco warrior with this 100% organic eco warrior slogan tshirt -

https://honestmiracle.com/collections/organic-slogan-t-shirts/products/eco-warrior-100-organic-t-shirt

Conclusion

I hope you've seen that you really don't need to be an extreme Instagrammer and reduce your yearly rubbish into one jar to help the planet and your health. If you can, then that's great, but for most of us that's overwhelming and when we feel overwhelmed we're not able to do as good a job.

It's important to go zero waste gradually, that way we avoid most of the overwhelm.

So, let me know what tips have helped you the most, I'm on Twitter https://www.twitter.com/kimcalera or on Facebook https://www.facebook.com/honestmiraclestore

If you need help with anything zero waste/eco/organic/toxin free you can email me and Emily at info@honestmiracle.com

Well done for caring that much about our planet that you got this book AND you made it to the end of this book! Did you know that puts you in the top 2% that are actually doing something to help our environment?!

You're doing a good job!

Lots of Green Love,

Kim xX

Printed in Poland
by Amazon Fulfillment
Poland Sp. z o.o., Wrocław

51412059R00063